Regina Célia Macêdo do Nascimento
Betânia C. Guilherme
Clemente Coelho Junior

The state of the art on the media's view of the mangrove ecosystem

Regina Célia Macêdo do Nascimento
Betânia C. Guilherme
Clemente Coelho Junior

The state of the art on the media's view of the mangrove ecosystem

Imprint

Any brand names and product names mentioned in this book are subject to trademark, brand or patent protection and are trademarks or registered trademarks of their respective holders. The use of brand names, product names, common names, trade names, product descriptions etc. even without a particular marking in this work is in no way to be construed to mean that such names may be regarded as unrestricted in respect of trademark and brand protection legislation and could thus be used by anyone.

Cover image: www.ingimage.com

This book is a translation from the original published under ISBN 978-613-9-70757-7.

Publisher:
Sciencia Scripts
is a trademark of
Dodo Books Indian Ocean Ltd. and OmniScriptum S.R.L publishing group

120 High Road, East Finchley, London, N2 9ED, United Kingdom
Str. Armeneasca 28/1, office 1, Chisinau MD-2012, Republic of Moldova, Europe
Printed at: see last page
ISBN: 978-620-7-25176-6

CHAPTER 1 GENERAL INTRODUCTION

The mangrove ecosystem is a transitional environment between land and sea, which is established in areas under the continuous action of the tides (OLMOS and SILVA, 2003). They are one of the richest environments on the planet and also one of the most fragile in the face of anthropogenic action, representing vitality for various species of birds, molluscs, fish, crustaceans and others. Mangroves have a high yield of fish and represent the basis of subsistence for a large part of the populations that live around the ecosystem (ARAÚJO *et al.* 2005).

Mangroves are subject to a number of adverse factors, such as poorly consolidated (muddy) soils, high salinity levels and poor aeration. These factors explain the low number of occurrences of plant species and the varied adaptations necessary for their survival (BASÍLIO *et al.* 2005). In Brazil, mangroves occupy a fairly significant area of the coastline, stretching from the state of Amapá to Santa Catarina (SCHAEFFER-NOVELLI, 1995).

There are essentially four types of mangrove in north-eastern Brazil: *Rhizophora mangle* (red mangrove) - a species located in a predominantly flooded area; *Laguncularia racemosa* (white mangrove) - preferably located in the interior of the mangrove forest, just after the *Rhizophora* strip; *Avicennia schaueriana* (black mangrove) and *Avicennia germinans* - these species are resistant to regions with a higher salinity index and are located in areas where the tide eventually comes in, already approaching the sandbank (SCHAEFFER-NOVELLI, 1995; VANNUCCI, 2002).

Even with all the productivity and integral protection provided for by federal law, the ecosystem has not been respected. There have been reports of intense degradation processes compromising its ecosystem services (BASÍLIO *et al.* 2005).

Deforestation, landfill, rubbish dumping, exposure to domestic and industrial effluents, dredging with direct and indirect damage, damming, predatory fishing are some of the problems present in mangroves (FERNANDES, 2012).

Based on this premise of destruction and devaluation of the ecosystem, environmental education, restoration and protection actions are being carried out as a counterpoint. It is therefore interesting to see what the media is reporting about this ecosystem by subsidising the state of the art.

The term "state of the art" is a translation from the English language, with the aim of surveying what is understood about a subject based on research carried out in a particular area (ROMANOWSKI and ENS, 2006). A study of the media's view of mangroves therefore contributes to our knowledge of how society conveys this information. Thus, the State of the Art is "an exposition of the level of knowledge and development of a field or issue" (SPINK, 1996 *apud* REIGOTA, 2007). From this perspective, the focus of the research is essentially exploratory, based on documents in the public domain.

In order to obtain the data needed to carry out the study, media content was analysed as a channel for influencing and shaping public opinion, addressing the processes of mangrove destruction, protection and preservation, recognising the importance of raising discussion about environmental education actions.

The exploratory research characterised a problem focused on the media's representation of environmental issues, especially the mangrove ecosystem, during a period from 2002 to 2014, due to the concentration of major destruction of mangrove areas for real estate development and other ventures.

As the study carried out was fundamentally exploratory, based on public domain documents, three distinct moments were observed. One related to the first contact with the investigative document, then the exploration of the material, unravelling aspects of the document, highlighting its most relevant points, and finally the categorisation and interpretation of the data obtained in the analysis.

CHAPTER 2 MANGROVE ECOSYSTEM: CONCEPT AND CHARACTERISTICS

The mangrove ecosystem is considered a transitional environment between the maritime and terrestrial regions (SCHAEFFER-NOVELLI, 1995; VANUCCI, 2002; OLMOS and SILVA, 2003; CORREIA and SOVIERZOSKI, 2005; FERNANDES, 2012).

Living beings and their non-living environment are interrelated and interact with each other. An ecosystem is any ecological unit or system that encompasses all the organisms that function together in a certain area, interacting with the physical environment in such a way that a flow of energy produces succinctly defined biotic structures and a cycling of materials between the living and non-living parts (ODUM, 1988).

The definition of a mangrove ecosystem also takes into account its hydrodynamic, faunal and floral characteristics, as well as its ecosystem dynamism. Due to the complexity of defining an ecosystem, Vannucci (2002) states that:

> Between the mangrove ecosystem and adjacent ecosystems. What actually unites the structure of the ecosystem into a coherent and functional whole is the dynamic interaction of its different parts, expressed as the transfer or flow of matter and energy from one component - or part - to the other components within the ecosystem and the dynamics of any ecosystem is extremely complex and always difficult to understand in every detail. (p.76)

Taking into account the general characteristics of mangroves, it is considered to be a transitional ecosystem with a marine and terrestrial environment, present in tropical and subtropical regions of the world. It is made up of woody plants called mangroves, with a characteristic algal flora associated with the ecosystem, making the environment favourable for reproduction, feeding and protection of the mangrove's characteristic animals (BARROS *et al*, 2000). Olmos and Silva (2003) consider this description and add the influence of the tides on the ecosystem to their definition.

When location is taken into account, mangroves are considered to be coastal forests that develop in the intertropical regions of the planet, in areas delimited by the influence of the tides, in sheltered areas, along estuaries, bays, deltas, inlets, river mouths and coastal recesses, inland brackish waters and lagoons, where river and sea waters meet (POR, 1989; VANNUCCI, 2002). Mangrove forests have a circumtropical distribution, with the greatest development in the area between the Tropic of Cancer and Capricorn (23°27'N and 23°27'S) (YOKOYA, 1995). In regions close to the equator, mangroves have exuberant vegetation, occupying extensive areas (CORREIA and SOVIERZOSKI, 2005).

In Brazil, mangroves occupy a very significant area of the coastline, stretching from Oiapoque in the state of Amapá (4°30' N) to Laguna in Santa Catarina (28°53' S), characterising around 1.3 million hectares according to data from the Chico Mendes Institute for Biodiversity Conservation (ICMBio) (SCHAEFFER-NOVELLI, 1995). The occupation of the area coincides with the stretch of coastline where the influence of the cold waters of the Falklands Current is small, running in a south-north direction along the coast of South America, presenting the second largest expanse of mangroves in the world (13,400 km^2), behind Indonesia (42,550 km^2) (OLMOS and

SILVA, 2003). Through studies related to remote sensing and local surveys, the state of Paraíba has around 97 km² of mangrove areas and Pernambuco around 162 km² (MAIA *et al.*, 2005).

In Brazilian coastal ecosystems, mangroves, coral reefs and estuaries have the greatest potential for renewable resources, with mangroves being the most productive in nature (DIEGUES, 1996).

The factors that determine dynamic changes in an aquatic ecosystem are hydrological, climatological, biological or mixed, when they work together (BARROS *et al.*, 2000). Therefore, for the mangrove to develop well, it is important to have an average temperature above 20°C and rainfall above 1,550 mm/year, without too many prolonged droughts (FERNANDES and PERIA, 1995). In the mangrove ecosystem, the flow of fresh water varies greatly and the conditions in this environment change with the variation in the volume of water carried. Obviously, if there is a gradual dilution of seawater, the biota must be prepared for these variations (BARROS *et al.*, 2000).

The tides are the main source of salt water entering the mangroves and are responsible for the variation in salinity in this environment. The greater the distance at which salt water penetrates the estuary will determine the limit of the mangrove towards the mainland. The amplitude of the tide will also determine the refinement of surface and interstitial waters, whose importance lies in the transport of propagules and nutrients, oxygenation, and the dispersal of larvae, especially of sessile animals (CORREIA and SOVIERZOSKI, 2005).

As the tides are limiting factors in the occurrence of mangroves, in some cases we see an area called apicum or salt marsh in the mangroves, located in their innermost part, in the mid- to upper-coast region. Its limits are the average levels of the sizigia tides and the equinoctial tides (SCHAEFFER-NOVELLI, 2015).

In addition to the characteristics mentioned by the authors Por (1989), Vannucci (2002), Fernandes (2012), Olmos and Silva (2003), Fernandes and Peria (1995), the mangrove is a tropical coastal environment with a differentiated flora that has had to adapt to a muddy substrate with great variations in salinity. For Schaeffer-Novelli (1995) mangroves are:

> Coastal ecosystem, a transition between terrestrial and marine environments, characteristic of tropical and subtropical regions, subject to the tidal regime. It is made up of typical woody plant species (angiosperms), as well as micro and macroalgae (cryptogams), adapted to fluctuating salinity and characterised by colonising muddy sediments with low oxygen levels. It occurs in sheltered coastal regions and provides favourable conditions for feeding, protecting and reproducing many animal species, and is considered an important transformer of nutrients into organic matter and generator of goods and services. (p.7)

Mangroves form an extremely important unit of fauna and flora, represented by typical vegetation and animals, with exclusive species associated with the ecosystem, some of which are of economic importance and of interest to public health (VANNUCCI, 2002).

Its flora is made up of typical woody, amphibious and evergreen plants, commonly known as mangroves (ARAÚJO and MACIEL, 1979). Mangrove is a word used to designate a species or group of species of trees and shrubs that have adaptations that allow them to colonise flooded land subject to the introduction of salt water (CINTRÓN and SCHAEFFER-NOVELLI, 1983; ADAIME, 1987). Mangrove species grow very well in fresh water, so the presence of salt makes them facultative halophytes, excluding other competitors (OLMOS and SILVA, 2003).

In the states of Alagoas, Paraíba and Pernambuco you can find essentially four types of mangrove:

- *Rhizophora mangle* (red mangrove) - a species that symbolises the mangrove because of its characteristic roots, located mainly on the banks of the estuary, a predominantly flooded area. Its roots act as ultrafilters that prevent salt from entering, as well as having lenticels that make it easier to breathe at high tides. Its propagules are the largest among the other mangrove species and can be transported for up to two years. Its leaves are shiny, while the inner part is more opaque; its bark is greyish and thin, and the wood is reddish due to the tannin;
- *Laguncularia racemosa* (white mangrove) - preferably located in the interior of the mangrove forest, just after the *Rhizophora* strip. They have a reddish petiole and vestigial glands. They have the largest flowering and propagule formation (racemes), which are generally 1 to 1.5 cm long. Most of the time we find this species growing in areas where fine sediment is deposited, where the current accumulates material;
- *Avicennia schaueriana* (black mangrove) and *Avicennia germinans* - these species are resistant to regions with higher salinity levels and are located in areas where the tide eventually

comes in, already approaching the sandbank. Its roots are buried horizontally, producing small breathing structures (pneumatophores) that have access to the surface, supplying oxygen to the root system and helping the plant to stick to the muddy surface. Its leaves are often covered in a layer of salt, due to the action of excretory glands on their surface (SCHAEFFER-NOVELLI, 1995; VANNUCCI, 2002; OLMOS e SILVA, 2003) (figure 2).

In addition to the four types of mangrove, we can find species associated with this ecosystem such as grassy and shrubby plants, mainly from the genera *Spartina*, *Hibiscus* and *Acrostichum*; cryptogams from the group Chlorophyceae, Phaeophyceae and Rhodophyceae, as well as phanerogams from the genus *Salicornia*, *Sesuviam* and the Chaenopodiaceae family (FERNANDES, 2012).

Because mangroves are transient areas, they are places where different faunas meet. The fauna that comes to inhabit mangroves is from terrestrial, marine and freshwater environments, animals that spend all or part of their lives in this ecosystem. The environment even imposes restrictions on the growth of organisms due to salinity, temperature, tidal influence and high turbidity, forcing the animals to adapt both morphologically and physiologically. After this period of adaptation to the mangrove, they will enjoy an environment abundant in nutrients, food and with a lower predation rate. In a general survey, the fauna of the Brazilian mangrove swamp showed around 60 species of crustaceans (crabs, shrimps, etc.), 35 of molluscs (snails, shellfish, etc.). 190 fish and 90 birds (including migratory birds) (FERNANDES, 2012; OLMOS and SILVA, 2003; SCHAEFFER-

NOVELLI, 2003).

Mangroves can have a fauna divided into the main communities:

1. **Tidal channels**: a predominantly flooded area, mainly containing the following groups: Fish: carapeba (*Diapterus* sp. *Eugerres* sp.), mullet (*Mugil* spp.); Crustaceans: freshwater and saltwater prawns (*Macrobrachium* sp., *Penaeus spp.*), Siris (*Callinectes spp.*); and plankton of marine origin.

2. **Margin of tidal channels**: area uncovered during low tides. Its main groups are: Decapod crustaceans: Siris (*Callinectes* spp.), freshwater and saltwater shrimps (*Macrobrachium* sp., *Penaeus spp.*) and mainly crabs of the *Uca spp.* genus; Bivalve molluscs: (*Venus spp.*, *Anomalocardia brasiliana* and *Crassostrea spp.*); and various birds, mainly boobies, herons, gulls and hawks that feed on the animals exposed by the tide.

3. **Woodland base**: an area located under the tops of trees and mangrove roots, where the substrate is no longer muddy but rigid. The main groups are: decapod crustaceans, particularly *Ucides cordatus*, *Goniopsis cruentata*, *Cardisoma* spp.; gastropod molluscs such as *Neritina spp.*, *Bulla spp., and* bivalve *Mytella spp.*; and in the largest preserved regions, reptiles and mammals.

4. **Trunks and aerial roots**: zone occupied by barnacles and oysters (*Crassostrea* spp.), gastropods (*Littorina angulifera*) and a flora of associated algae and lichens growing on the trunks, branches and roots.

5. **Tree tops**: they represent an area of interaction between the

marine and terrestrial environments. They contain insects such as termites, ants and crickets; birds such as herons and egrets nest in the trees, while grebes, hawks and gulls, among others, use them for hunting; species of snakes and some amphibians also occur in this environment (MAIA *et al.*, 2005).

Mangrove areas have been ecologically and socio-environmentally important since ancient times, due to the factors mentioned above. The exploitation of Brazilian mangroves in prehistoric times has been theoretically demonstrated through the analysis of sambaquis scattered along the Brazilian coast, showing the fauna associated with mangroves by prehistoric nomadic tribes (VANNUCCI, 2002).

In addition to the mangrove serving as a source of food and income for the communities that live or work in its surroundings (FERNANDES, 2002), mangrove trees serve to protect the coastline from erosion, acting as a barrier and reducing the action of the tides on the sediment (COELHO JR and NOVELLI, 2000).

Given the great diversity of functions and ecosystem services provided by mangroves, the following tasks stand out: source of particulate and dissolved organic matter for the neighbouring coastal waters, forming the base of the trophic chain with species that are economically and ecologically important; area for shelter, reproduction, development and feeding of different species; protection of the coastline by combating erosion; flood prevention and protection against storms; filtration of sediments and pollutants; source of recreation and leisure, related to its landscape and scenic value; source of various foods, which is associated with the community that lives around this mangrove area. As such, the mangrove is considered a nursery for life, especially marine life (COELHO JR and NOVELLI, 2000).

Despite the importance of the mangrove for ecological and human equilibrium, what we see is the misuse of the ecosystem in coastal areas, putting it at risk and destroying it through urban-industrial occupation processes, causing predatory exploitation of the fauna and flora and pollution of its waters, as well as turning it into rubbish dumps and landfills (CUNHA, 2000; OLIVEIRA, 2004; ROCHA *et al.*, 2006). As a result of this misuse, the mangroves are associated with rubbish and stench (ALMEIDA, 2010). These actions have a variety of consequences, including a reduction in primary, estuarine and coastal productivity and, consequently, a reduction in coastal fishing production, the source of food and income for riverine populations (LIRA *et al.*, 1992).

The media can be used both for and against a situation. When issues are publicised and encouraged by the media, they can take on enormous proportions, bringing about a result beyond what was expected. The existence of a democratic and open media promotes a sustainable development model for society (FREY, 2002).

> [...] in the society established from the 20th century onwards, the media began to play a central role in the dissemination of information. The media became part of people's day-to-day lives and gained an important place in the political sphere, as it is a space for visibility (MASSUCHIN, 2010, p.320).

There is a divergence between the media and the environment: at times they play a critical role in developing environmental issues, while at other times they are responsible for omitting important events related to the environment (RAMOS, 1995). About 30 years after the end of the 20th century and in the early years of the 21st century, the environment has gained prominence on the world economic, political and intellectual stage. Discussions on issues related to the environment were highlighted, taking globalisation into account (HISSA, 2008).

The media's affiliation with environmental issues was chosen because of the growth of science journalism. If there hadn't been this advance in the specialisation and qualification of professionals working in this area, the development and production of environmental news and controversies would have continued at a reduced standard (ANDRADE, 2007). Studies on the media and the environment have grown considerably in recent decades. Possibly, the politicisation of the environmental discussion has also been ensured by the solidification of the communications sector and the expansion of the cultural industry around the world (ALMEIDA JR. and ANDRADE, 2007). Even with the increase in publications and media development, the training and entertainment industry has accumulated control over a large part of the content broadcast around the world, including television, radio and newspapers, influencing the behaviour, consumption patterns and habits of those who receive the content (TRIGUEIRO, 2003).

From the 1990s onwards, as the process of development and globalisation changed politics and cultural practices, there was a trend towards a crisis in the traditional media and the expansion of electronic language, reshaping the formats of communication companies, the production of messages and reports, as well as consumer practices (ORTIZ, 1994 *apud* ALMEIDA JR. and ANDRADE, 2007).

The written or printed media is the second largest vehicle for disseminating journalism, second only to television (TRIGUEIRO, 2003). The appreciation of various genres of journalistic material published in print and digital media provides readers with knowledge of the quality of life of human beings in society (FERNANDES, 2015).

The press emerged with the importance of disseminating information and specialising in the task of informing through writing. Thus, journalism came to occupy a key role and consequently has news as its main product (VIANA, 2013). As for the specificity and form of the written media, the author states:

> Print media stands out first and foremost for being a device that is largely geared towards journalistic and/or advertising

material, printed in printers using a specific technology. This printing technology has been improved since the invention of the printing press by Johannes Gutemberg in the 15th century, which made mass printing possible. (p. 2)

The media's use of language suited to the level of understanding of the masses has meant that communication has started to provide contributions so that the population can see itself in an evaluative understanding of its past, trajectory, development and the projection of its future (FERNANDES, 2015).

The journalistic environment still has a fragmented view of environmental issues, resulting from professionals who do not have specific training in environmental issues, resulting in editorials with no "defined space" for the environment or for helping to raise ecological awareness (JONH, 2001). The written media's treatment of environmental issues is associated with denunciations, problems related to ecological niches and finally, still absent from the media agenda, solutions to environmental issues (OLIVEIRA, 1996).

Despite this, the media are responsible for disseminating issues related to environmental problems, becoming "an *essential element in finding ways to resolve conflicts of political and economic interests, becoming a limiting factor in achieving a globalising vision of the environment*" (FERNANDES, 2015).

ART

To describe the state of the art on the media's view of the mangrove ecosystem in the states of Alagoas, Paraíba and Pernambuco. With specific reference to:

- Diagnose how the media participates in building public opinion on events related to the ecosystem;

- To analyse the journalistic material (articles and notes) published about the mangrove;

- Evaluate the published content, defining the greatest incidences of degradation, protection, preservation and environmental education of the mangrove ecosystem in the different states.

CHAPTER 5 RESEARCH PROCEDURES

In order to establish the relationships between society, the press and mangroves, the content was analysed through a digital bibliographic survey of the main newspapers in the states of Alagoas, Paraíba and Pernambuco from 2002 to 2018. The states were chosen because of their similar flora and the growing economic importance of mangrove areas. The most representative media in circulation are: Alagoas (Alagoas 24h, Gazeta de Alagoas and Primeira Edição), Paraíba (Correio da Paraíba, Globo and Jornal da Paraíba) and Pernambuco (Diário de Pernambuco, Folha de Pernambuco and Jornal do Comércio).

As for content analysis, according to Silva (2011) it can be defined as "*a methodological principle applicable in qualitative research during the process of evaluating and interpreting the material collected*".

In order to map out how such news is reported in the media, three categories were chosen, taking into account the following aspects:

a. **Social**: when it involves issues relating to society, such as benefits for both sides, issues involving community participation and even peculiar cases such as homicides;
b. **Economic**: when the news revolves around the construction and industrialisation process;
c. **Government**: when the news places organisations government (federal, state or municipal) as precursors of the process;

The following aspects were taken into account when analysing the data:

a) Number of articles published on mangroves;

b) Content of the articles and their scope and focus (environmental degradation, protection, preservation and education);

Analysis can be considered one of the most important empirical research techniques in the field of social science, especially in matters related to journalism and communication (CARVALHO and TONI, 2009). News can also be classified as a type of generalist knowledge; however, it reinforces that news has its own place in terms of classification (PARK, 2002).The documentary analysis took place between 2002 and 2018, when there was a concentration of major property development, the destruction of mangrove areas for the growth of shrimp farming, among other developments.

Analyses were carried out at three different times:

I. **Pre-analysis**: the actual organisation, corresponding to the choice of documents to be included in the analysis, together with the formulation of hypotheses and objectives and the drawing up of a guide to help and support the final interpretation. Pre-analysis establishes contact with the document under investigation;

II. **Exploration of the material**: analysis proper, where the aspects to which the publication refers are broken down, listing the points and hypotheses raised in the pre-analysis;

III. **Treatment of the results obtained and interpretation**: using simple operations such as percentages to establish graphs, figures and tables of results. This included the categorisation already pre-defined in the screening (BARDIN, 2006, *apud* MOZATTO and GRZYBOVSKI, 2001).Given the above description, Mozatto and Grzybovski (2001) note that *"[...] content analysis is a set of techniques for analysing communications, which aims to overcome uncertainties and enrich the reading of the data collected"*. In order to successfully screen the materials, research processes were followed during this period, such as: analytical description, related to a kind of treatment of the information contained in the reports, and categorical analysis, which can be defined as a type of analysis that aims to take into account the totality of a text, passing it through the study of classification (SILVA, 2011).For a content diagnosis to be of value, there are some prerequisites, such as: the quality of the conceptual elaboration first made by the researcher, the precision with which it will be translated into modifiable, the analysis scheme or categories and, absolutely, the agreement between the reality to be analysed and these categories (FREITAS *et al,* 1997).The choice to analyse the content of the monograph is the most appropriate to study the messages issued by the written media where the term Mangrove Ecosystem could be identified. When content analysis is used as a support, the aim is to go beyond a simple reading, but rather to observe the conditions that determine the sender's stance when producing the report. Another highlight is the news published as a case study, when it refers to an intensive analysis of a particular situation (YIN, 2001 *apud* MOZATTO and GRZYBOVSKI, 2001).

From January 2002 to July 2018, the most influential written/digital press in the states of Alagoas, Paraíba and Pernambuco published 255 articles and opinion pieces related to the mangrove ecosystem. The story, also known as a report, is a piece of journalism of an informative nature, so it has an ascendancy because it has great potential to reach the largest number of readers through its approach, providing greater viability of content to the reader. The report can have several foundations, as it allows for a variety of reporting possibilities, as long as the authenticity of the report is not compromised (FUSER, 1996; SILVA 2011).

It was possible to see that between the years 2010 and 2014, since the selection of publications in journals, the volume of citations related to the Mangrove Ecosystem was more relevant, with 143 publications - 4 in 2010 (3%); 11 in 2011 (8%); 47 in 2012 (32%); 33 in 2013 (24%) and 48 last year (33%), corresponding to 56% of the total universe.

These results are backed up by the fact that in 2012 the media took a slightly stronger look at the environment, highlighting the mangrove swamp. This ecosystem was (still is) suffering from devastation, so there were educational campaigns to raise awareness of its preservation, along with government enforcement actions. In 2014, with the great demand for developments in the region, the mangrove suffered greatly from anthropic actions, with news of invasions and developments standing out. On the other hand, as was the case in 2012, actions were taken to defend it. In view of the time frame, there was a low demand for articles and notes related to the mangrove ecosystem, leaving environmental journalism wanting. Despite the low numbers, we have seen growth since 2010. Bueno (2007) says that a quick look at the media's repercussions on the environment reveals the need to conceive different realities and requirements in the process of journalistic production that is orientated towards environmental issues. Still in this context, Coletto *et al.* (2010) reported that coverage that definitively sets out to deal with facts linked to the environment needs to meet the requirements of the correct teachings and address the issue in a broad and contextualised way.As the research was carried out in three states, in order to see if there were any differences between the press reports on the subject, when observing and sorting the publications by the most influential media outlet in the states in its category (electronic newspaper), it was observed that Gazeta de Alagoas and Jornal do Comercio (PE) were the most dedicated to publishing about the mangrove. Both newspapers emphasised a greater social focus, with many of their stories focusing on the community's use of the ecosystem, both positively and negatively. After selecting the reports, it can be seen that the states of Alagoas and Pernambuco were more representative of the mangrove ecosystem than Paraíba.

Taking into account the aspect related to the journalistic manifestation, it was found that the social nature presented constant publications throughout the study period.

The reports were related to the regular and irregular use of the ecosystem by the population, such as "*Mangue é fonte de sobrevivência*" (Mangrove *is a source of survival)* and "*Famílias invadem área* de mangue" (*Families invade mangrove area),* both in the newspaper Gazeta de Alagoas.

Figura 1. Excerpts from reports on the use of mangroves by the population

THE PRESS' OWN DISCOURSE
Born and raised in Palatea, as she herself says, "Bastinha", the name by which she is known in the community, never misses an opportunity to recall the damage caused by the group that exploited the residents in an oyster production project. The people of Alagoas remember the orchestrated coup carried out by the Oceanus Institute, an NGO created with the specific purpose of stealing public money. Fishermen and shellfish gatherers from at least ten municipalities on the **Alagoas** *coast were duped by a group that promised sustainable development, monthly income, a better life [...] -* **Mangrove is a source of survival (Gazeta de Alagoas, 08/04/2012)**
Families calling themselves homeless occupied a sandbank area on the banks of the Persinunga River in the Peroba district of Maragogi, on the border between Alagoas and Pernambuco, last Sunday morning. The Municipal Environment Department received reports that the invaders had deforested the mangrove swamp and called in the Brazilian Institute for the Environment and Renewable Natural Resources (Ibama), which is going to carry out an inspection of the site [...] - **Families invade mangrove area (Gazeta de Alagoas, 15/07/2008)**

Society's misuse of mangroves is due to a lack of environmental awareness. Jacobi (2003) states that the dependent attitude

The lack of information, lack of environmental awareness and lack of community practices based on the participation and involvement of citizens, which insinuated a culture of rights based on motivation and complicity in environmental management.

Another relevant aspect that caught the eye when sorting social data was the use of the ecosystem as a police case. Especially in the state of Alagoas, many bodies were dumped in the mangroves during the period studied.

<table>
<tr><td>THE PRESS' OWN DISCOURSE</td></tr>
<tr><td>

On Monday afternoon (03) a fisherman found the body of a man in the waters of the Sanhauá River mangrove swamp in the town known as Ponto do Moinho, in the centre of Bayeux, in the metropolitan region of João Pessoa. According to information from ABS Fire Brigade VTR Lieutenant Roberto, who was called to the scene by the Military Police, the body was in a state of decomposition and had no identification. Lieutenant Roberto of the Fire Brigade's rescue team said that a boat was used to reach the place where the body was and that the search took seven minutes of sailing [...]- **Body found in a state of decomposition in the mangrove swamp in Bayeux (Correio da Paraíba, 03/04/2012)**

</td></tr>
<tr><td>

The murder of a 15-year-old teenager on the afternoon of Wednesday 19 shocked residents of the village of Cabreiras, in the municipality of Marechal Deodoro, for a futile reason: the exchange of a bicycle. According to information gathered at the scene, the 15-year-old teenager contacted his tormentor - a 17-year-old - to destroy a bicycle he had bought. In a rage, the accused took the victim to a mangrove area and strangled him. The boy's body was thrown into the mangrove swamp, in a place that is difficult to access and will need to be removed with the help of a team from the Fire Department [...] - **Teenager strangled and thrown into mangrove swamp (Alagoas 24H, 19/11/2014)**

In the early hours of Thursday 29th, the Paraíba Civil Police launched the 'Mangue Seco Operation' in the town of Pitimbú, on the south coast of Paraíba. In all, seven people were arrested, including two women. Also seized were four 38 calibre revolvers, three ballistic waistcoats, 48 rocks of crack, marijuana plants and seeds. According to police chief Aldroville Grisi, the operation began at 4.30am with the aim of dismantling drug trafficking in the municipality [...] - **Gang that trafficked on the banks of a river and mangrove swamp in Pitimbú arrested (Correio da Paraíba, 29/11/2012)**

</td></tr>
</table>

Figure 2 - Excerpts from reports on police cases

But this growth has a huge impact on the environment. Its large scale is only beneficial to the economy, leaving nature at a disadvantage. In the midst of this advance in construction, only one edition of the Gazeta de Alagoas newspaper was found in 2012 that mentioned ecological compensation: *"DER must replant mangrove vegetation"*. Another 7 news items featured sustainable ecological tourism or relatively correct exploitation. One that stands out is the article entitled *"Pontas de Mangue: Novos Encantos no Litoral Norte" (Mangrove tips: new charms on the north coast)* from the newspaper Gazeta de Alagoas, published in 2007.

THE PRESS' OWN DISCOURSE
Maragogi, AL - If Maragogi is paradise within reach on the north coast of Alagoas, the village of Ponta de Mangue is the most beautiful in the region. The place still has deserted beaches and the air of an authentic fishing village. For tourists, there are accommodation options ranging from the refinement of a resort to the simplicity of a camping area [...] - **Pontas de Mangue: New Charms on the North Coast (Gazeta de Alagoas, 05/01/2007)**

CHAPTER 9 MEDIA APPROACHES TO THE GOVERNMENTAL

With Brazil developing, it is a fact that the government must turn its attention to acclimatising its natural resources. The news was related to attitudes aimed at protecting mangrove areas, such as the region's characteristic animals. Cavalcanti (1997) states that it is essential to add ecological (or eco-social) concerns to public policies in Brazil.

This ecological concern associated with the mangrove ecosystem was highlighted in the article entitled *"City Hall starts cleaning up Cabanga's mangroves"*.

Figura 4. Excerpt from a report on ecological concerns

THE PRESS' OWN DISCOURSE
This Wednesday (21) and until next Saturday (24), a joint clean-up effort began in the Cabanga mangrove swamp, managed by Recife City Hall and involving 15 employees from the Maintenance and Urban Cleaning Company (Emlurb). During the three days of cleaning, the City Hall expects to collect more than 20 tonnes of rubbish from the mangrove. According to Emlurb's Director of Urban Cleaning, Rodrigo Braynner, all the work will be carried out manually so as not to damage the vegetation [...] - **City Hall starts cleaning Cabanga mangrove swamp (Jornal do Comércio, 21/09/2011)**

Unfortunately, over the years, the demand for mangrove areas has only increased, which favours an imbalance in the ecosystem. On the other hand, political representatives are trying to preserve the area, which is very rich for humans as a source of income, food and housing.

Among the topics covered in the research, environmental degradation was the second most prominent. Tavares and Freire (2003) colour environmental information as a type of scientific and technological information of fundamental importance in overcoming the environmental crisis we are currently experiencing, contributing to the preservation of natural environments and those incorporated by man. As the written media is one of the most influential in shaping a population's critical awareness, it is of the utmost importance that it exposes environmental events.

Issues relating to environmental degradation were all found in publications related to the economic factor. As entrepreneurs use the mangrove area to build houses, resorts, shrimp farms and so on, the ecosystem is losing its area in the midst of globalisation. Some of the reports that highlight these facts are those in the Gazeta de Alagoas newspaper entitled "*Mangrove is landfilled in Trapiche to build houses*" and "*Village is built in mangrove area*" (figure 5).

The news didn't delve deeply into the issue, without formulating a supposed solution. Alves (2002) emphasised that most media outlets conflict ideas of environmental protection and respect.

As of yesterday, the Municipal Secretariat for Environmental Protection (SEMPMA) had not identified the person responsible for the landfill of a mangrove area in the locality known as Sítio Recreio, in Trapiche da Barra. Last Saturday, technicians from the Environment Institute (IMA) spotted buckets dumping sand and clay on the site and were due to return to embargo the work. But it was Sempma technicians who ended up inspecting the area yesterday morning. According to Sempma inspector Paulo Nunes, the site will be monitored over the next few days to prevent the landfill from continuing. [...] - **Mangrove is landfilled in Trapiche to build houses (Gazeta de Alagoas, 13/09/2005)**

The Persinunga River is the natural landmark that separates LA from PE at the southernmost tip of the coast. Over the years, however, the watercourse has suffered from numerous environmental aggressions. The biggest is the irregular occupation of its banks. Now, an entire village is being built within a Permanent Preservation Area (APP), on the banks of the river, which is part of the APA Costa dos Corais Conservation Unit. The invasion has already led to the suppression of mangroves and the landfill of the right bank. According to Gazeta, the occupation began around two years ago, but has recently intensified [...] - **Village is built in mangrove area (Gazeta de Alagoas, 01/11/2012)**

Figura 5. Excerpts from reports on environmental degradation in the economic sector

With global development on a large scale, many news stories about environmental impacts aren't attracting as much attention from the population, and consequently the media doesn't pay much attention to them. Trigueiro (2003) reinforces this hypothesis by questioning the immediacy of the population, which gives less and less value to what will happen in a few years' time. With this reaction, we can see in some publications related to the economy and degradation in which mangrove areas are constantly being landfilled for the construction of major works or in the more urban environment, or in more beach-orientated areas for the construction of luxury hotels.

Among the news items aimed at society and the government, newsletters related to educational actions to raise awareness of the importance of preserving the mangrove, in view of its biological richness, such as the publication in the Diário de Pernambuco, published in 2012, entitled *"Campaign raises awareness among residents about the importance of the mangrove"* (figure 6).

THE PRESS'S OWN DISCOURSE
In the spirit of Christmas, children from the municipality of Paulista are taking part in an ecological walk this Sunday (23rd) at 9am in the Timbo river estuary. Organised for 13 years by the Pau Amarelo Environmental Preservation Group (GPAPA), the Father Christmas in the Mangrove campaign provides information about the importance of mangroves and promotes symbolic cleaning of the area. At the end of the event, the participating children, who are fishermen's children, will receive presents from Santa Claus. - **Campaign raises awareness among residents about the importance of mangroves (Diário de Pernambuco, 23/12/2012)**

Figura 6. Excerpt from report related to environmental education action

Carrying out activities aimed at the environmental context, both in terms of improving it and preserving it, is highlighted by Jacobi (2003) when he emphasises that "*reflection on social practices, in a context marked by the permanent degradation of the environment and its ecosystem, involves a necessary articulation with the production of meanings about environmental education*".

As for the publications related to preservation and protection, some highlighted the closed season and the rescue of individuals of the *Ucides cordatus* species, an animal that plays an extremely important role in the mangroves, as well as generating income for many families who make a living from the ecosystem (figure 7). According to Nordi (1994), *Ucides cordatus* basically spends the whole year holed up in its individual galleries, which can be up to 1m deep, under the mangrove trees. It is caught at low tide by fishermen who often use only their hands, and is of paramount importance for the subsistence of the local population.

News like this attracts readers' attention, leading to a greater interest in delving deeper into environmental issues. According to Ferreira *et al.* (2005), when there is a change in the environment, whether to a greater or lesser extent, as a result of human action, it can affect the health of the population, endanger animal species and compromise environmental quality.

THE PRESS'S OWN DISCOURSE

This Tuesday (12) marks the start of a ban on catching the mangrove crab in Pernambuco. This is due to the reproduction period of the crustacean, which is commercialised for culinary and craft purposes. Those who sell the crab must declare the quantity they have in stock to Ibama in order to be able to trade legally [...] - **Ban on catching mangrove crab in Pernambuco begins (Diário de Pernambuco, 13/203/2013)**

More than 1,500 mangrove crabs were rescued by the Environmental Police Battalion on Thursday night (18) in the rural area of Santa Rita, in Greater João Pessoa. The animals were being transported in burlap sacks in a vehicle. According to the police, the crabs had been caught using a net, a practice prohibited by environmental legislation. Major Magno Fonseca explained that the rescue of the animals took place

during a routine approach to the vehicle [...] - **In PB, more than 1,500 mangrove crabs are rescued by the Environmental Police (Jornal da Paraíba, 19/12/2014)**

Figure 6 - Excerpts from reports on *Ucides cordatus*

We can see the approach to environmental issues as something that "[...] *allows us to interpret the essence of a global media interface; therefore, analysing the media allows us to detect the important role it plays in disseminating national and international public policies in this area. Certainly, this highlights its opinion-forming character*" (MENEZES, 2011).

However, the analysis of the state of the art on the media's view of the mangrove ecosystem, in the screening of 17 years of media content, proved to be insufficient to promote any change in critical thinking or the construction of a public opinion of society related to the ecosystem, It was noted that there was a low number of publications and depth in their articles, many of which did not investigate the causes of environmental destruction, for example, and consequently did not generate solutions to the problems exposed.

The media outlets Gazeta de Alagoas, Jornal do Comércio and Diário de Pernambuco are contributing more and more to environmental reporting, whether it's on degradation, environmental education or protection and preservation. On the other hand, the state of Paraíba appears with very few publications, totalling around 10% of the overall universe. Despite the low numbers, its news was diverse, including governmental, social and economic issues.

Social reports are taking up more and more space in the media, and their gradual increase confirms this, stimulating greater interest among the target audience, resulting in greater proximity and identification with environmental issues. The closer people get to the environment, the more likely they are to take action to preserve it. In the case of the mangrove swamp, an ecosystem that generates income for many people, especially those living along the river banks, the population comes to understand that it is not a place for rubbish.

Most of the mangrove-related publications were presented clearly and succinctly to readers, making it easier for them to understand the situations reported and enabling them to adopt a cohesive approach to the facts.

In general, the articles were attractive to the public, providing some curiosities about the object of study, leisure tips and ecological tourism. On the other hand, the media outlets studied left something to be desired in terms of their investigative approach to mangrove problems, without presenting solutions to the problems reported in their editions.

LIST OF THE STATE OF THE ART

TITLE	JOURNAL	YEA	CLASSIFICATION
COMPLEX POLLUTES MANGROVE AREA IN MARAGOGI	GAZETTE	2002	SOCIAL
CONSTRUCTION THREATENS MANGROVES	DIÁRIO PE	2002	ECONOMIC
DEPOT THAT FILLED IN MANGROVE AREA IN MUNDAÚ DEMOLISHED	GAZETTE	2002	ECONOMIC
IMA FINES LANDOWNER FOR CLEARING MANGROVES	GAZETTE	2002	GOVERNMENT
MANGUE VERDE INSTITUTE AND OAM SIGN AGREEMENT	GAZETTE	2002	GOVERNMENT
MANGROVE IS DESTROYED IN PRATAGY	GAZETTE	2002	SOCIAL
LOOKS LIKE A CARTOON	DIÁRIO PE	2002	SOCIAL
THERMAL POWER PLANT SUSPECTED OF ENVIRONMENTAL DAMAGE	DIÁRIO PE	2002	GOVERNMENT
MANGROVE OBSERVATION TOWER	DIÁRIO PE	2002	SOCIAL
COMPANY FINED FOR DEFORESTING MANGROVES	GAZETTE	2003	ECONOMIC
INADEQUATE OYSTER EXTRACTION THREATENS MANGROVES	GAZETTE	2003	ECONOMIC
CRAB DISAPPEARS, WORSENING MANGROVE ECONOMY	GAZETTE	2004	ECONOMIC
MANGROVE AREA LANDFILLED TO BUILD HOUSES IS BANNED	GAZETTE	2005	ECONOMIC
CONSTRUCTIONS LANDFILL MANGROVE SWAMP IN MUNDAÚ	GAZETTE	2005	ECONOMIC
ENFORCEMENT LOSES ENVIRONMENTAL WAR	GAZETTE	2005	GOVERNMENT
GROUP PRESSES FOR WORK IN MANGROVE SWAMP	GAZETTE	2005	ECONOMIC
MANGROVE SWAMP LANDFILLED AT TRAPICHE FOR HOUSING DEVELOPMENT	GAZETTE	2005	ECONOMIC
MANGROVES BECOME RUBBISH IN THE SANTA RITA APA	GAZETTE	2005	SOCIAL
CITY HALL ORDERS MANGROVE LANDFILL	GAZETTE	2005	GOVERNMENT
INCORRECT RUBBISH DISPOSAL HARMS FISHING IN MANGROVES	PB JOURNAL	2006	SOCIAL
DESTRUCTION IN MANGROVE AREAS DESTROYS HALF OF SANCTUARY	GAZETTE	2006	ECONOMIC
THE MANGROVE MEETS THE SEA IN SAUAÇUHY	GAZETTE	2006	SOCIAL

FOREIGNERS DISCOVER THE NORTH COAST - PONTA MANGUE	GAZETTE	2006	ECONOMIC
NEW MUD IN THE MANGROVES	GAZETTE	2006	SOCIAL
MANGROVES SURPRISE CHILDREN	GAZETTE	2006	SOCIAL
OPERATION MANGUEZAL UNCOVERS CAR DISMANTLING BASE	24H	2006	GOVERNMENT
GETTING TO KNOW THE MANGROVES	GAZETTE	2007	SOCIAL
CRAB EXPLOITATION GETS OUT OF HAND	GAZETTE	2007	SOCIAL
THE BRAIN OF THE MANGROVE	GAZETTE	2007	SOCIAL
PONTA DE MANGUE: NEW CHARMS	GAZETTE	2007	ECONOMIC
STEELWORKS DESTROY MANGROVE SWAMP	GAZETTE	2007	ECONOMIC
THE MANGROVE REVOLT	GAZETTE	2008	SOCIAL
REGIONAL MEETING ON ENVIRONMENTAL EDUCATION IN MANGROVE AREAS	24H	2008	SOCIAL
FAMILIES INVADE THE MANGROVE SWAMP	GAZETTE	2008	SOCIAL
BOY KILLED AND DUMPED IN MANGROVE SWAMP	24H	2008	SOCIAL
RIVERSIDE MANGROVE	DIÁRIO PE	2008	GOVERNMENT
CATAMARAN RIDE IN THE PRESERVATION INITIATIVE	DIÁRIO PE	2008	SOCIAL
JOURNEY TO THE CENTRE OF THE MANGROVE	GAZETTE	2008	SOCIAL
ENVIRONMENTAL COMPLIANCE ON THE PORTO-MARACAÍPE MOTORWAY	DIÁRIO PE	2009	GOVERNMENT
MPF WANTS TO DEMOLISH SHACKS ON THE BANKS OF A MANGROVE SWAMP	PB JOURNAL	2009	GOVERNMENT
OPERATION INSPECTS CONSTRUCTION IN MANGROVE AREAS	PB JOURNAL	2009	GOVERNMENT
POLLUTION IN THE CHICO SCIENCE MANGROVE	DIÁRIO PE	2009	SOCIAL
MANGROVE SUPPRESSION PLANNED	DIÁRIO PE	2009	GOVERNMENT
SEWAGE DESTROYS MANGROVES IN JP	PB JOURNAL	2010	SOCIAL
IBAMA: ORCHID AND MANGROVE EXHIBITIONS	24H	2010	SOCIAL
SINGLE-ENGINE PLANE CRASHES IN PINA MANGROVE SWAMP	DIÁRIO PE	2010	SOCIAL
PARIPUEIRA AND ITS MANGROVES	FIRST ED.	2010	ECONOMIC
THE TIDE IS IN THE HARBOUR	JC	2011	ECONOMIC
FIREFIGHTERS SUSPEND MANGROVE SEARCH	PB JOURNAL	2011	SOCIAL
A CORPSE IS FOUND IN A MANGROVE SWAMP IN CORURIPE	24H	2011	SOCIAL

Title	Source	Year	Category
WOMAN'S BODY FOUND IN MANGROVE SWAMP	PB JOURNAL	2011	SOCIAL
ENVIRONMENTAL EDUCATION FOR MANGROVE AREAS	24H	2011	SOCIAL
ALLIGATOR CAUGHT SUNBATHING	JC	2011	SOCIAL
MORE THAN 20 TONNES OF RUBBISH TO BE REMOVED FROM THE MANGROVES	DIÁRIO PE	2011	SOCIAL
MANGROVES IN IMBIRIBEIRA THREATENED BY IRREGULAR OCCUPATION	JC	2011	ECONOMIC
NO WORK ON THE MANGROVE ROAD	JC	2011	ECONOMIC
CITY HALL STARTS CLEANING UP CABANGA MANGROVE SWAMP	JC	2011	GOVERNMENT
MANGUABA RIVER AND ITS MANGROVES	FIRST ED.	2011	SOCIAL
THE STORY OF A PAPER PARK	JC	2012	SOCIAL
ENVIRONMENTALISTS MAKE SYMBOLIC ACT WARNING OF MANGROVE EXTINCTION	JC	2012	GOVERNMENT
HOME HEATING COMES FROM MANGROVE BACTERIA	FIRST ED.	2012	SOCIAL
MANGROVE AREAS OF THE PRATAGY RIVER ARE BEING USED FOR SHOWS	FIRST ED.	2012	ECONOMIC
THE BEAUTIFUL LANDSCAPES OF RECIFE THAT FEW KNOW ABOUT	JC	2012	SOCIAL
HEARING TURNS INTO PRO-SHIPYARD ACT	GAZETTE	2012	GOVERNMENT
SOLAR BOAT IN THE MANGROVES	JC	2012	SOCIAL
BATTALION NOTICES ENVIRONMENTAL DEVASTATION	GAZETTE	2012	GOVERNMENT
CAMPAIGN RAISES AWARENESS AMONG RESIDENTS ABOUT THE IMPORTANCE OF MANGROVES	DIÁRIO PE	2012	SOCIAL
CRAB PICKERS ARRESTED IN MANGROVE SWAMP	PB JOURNAL	2012	SOCIAL
COMPENSATION HAS ALREADY REPLANTED 1/3 OF PROMISED MANGROVES	FIRST ED.	2012	GOVERNMENT
ENVIRONMENTAL AWARENESS IN THE MANGROVE SWAMP	24H	2012	SOCIAL
BODY FOUND IN MANGROVE SWAMP IN BAYEUX	CORREIO PB	2012	SOCIAL

Title	Source	Year	Category
DER MUST REPLANT MANGROVE VEGETATION	GAZETTE	2012	ECONOMIC
MANGROVE AND ATLANTIC FOREST DEFORESTATION DISCOVERED IN PORTO CALVO	GAZETTE	2012	SOCIAL
MANGROVE DOCUMENTARY CONTENDS FOR AWARD	FIRST ED.	2012	GOVERNMENT
COMPANY STRUGGLES TO RESTORE MANGROVES	GAZETTE	2012	ECONOMIC
FAMILIES BUILD SHACKS IN MARACAÍPE MANGROVE SWAMP	DIÁRIO PE	2012	ECONOMIC
FERNANDO DE NORONHA EMBRACES MANGROVES	DIÁRIO PE	2012	SOCIAL
LARGE RESORT IN THE MIDDLE OF THE MANGROVES OF ALAGOAS	24H	2012	ECONOMIC
ENVIRONMENTAL IMPACT WOULD BE GREAT, SAYS IBAMA.	GAZETTE	2012	GOVERNMENT
YOUNG ENTREPRENEURS UNITED FOR A BIG LEAP	JC	2012	ECONOMIC
LANDFILL OF MANGROVE AREA	JC	2012	ECONOMIC
MANGROVES ARE A SOURCE OF SURVIVAL	GAZETTE	2012	SOCIAL
MANGROVES IN MARACAÍPE GIVE WAY TO BOATS	DIÁRIO PE	2012	ECONOMIC
MANGROVES MAKE A DIFFERENCE	FIRST ED.	2012	SOCIAL
MANGUE MAKES A DIFFERENCE, MOBILISES HUNDREDS OF PEOPLE ON THE WATERFRONT.	FIRST ED.	2012	SOCIAL
MANGROVES IN ALAGOAS CALL FOR HELP	GAZETTE	2012	SOCIAL
SUAUÇUHY MANGROVES	FIRST ED.	2012	SOCIAL
CAPIBARIBE RIVER MANGROVES UNDERGO JOINT EFFORT	DIÁRIO PE	2012	GOVERNMENT
MANGROVES IN CATUAMA	DIÁRIO PE	2012	SOCIAL
MANGROVES IN DANGER	DIÁRIO PE	2012	SOCIAL
NAVY WON'T GIVE UP SELLING MANGROVES TO BUILD PARK	JC	2012	ECONOMIC
MISERY IN SIGHT IN THE PARK AREA	JC	2012	GOVERNMENT
MPF AND IBAMA PRODUCE DOCUMENTARY ON MANGROVES	24H	2012	GOVERNMENT
MPF WANTS TO PREVENT POSSIBLE DISASTER IN MANGEUZAL AREA	JC	2012	GOVERNMENT

Title	Source	Year	Category
PINA JOINS THE ROUTE	JC	2012	ECONOMIC
PM FINDS BODY OF MISSING YOUNG MAN IN MANGROVE SWAMP.	CORREIO PB	2012	SOCIAL
POLLUTION PUTS MANGROVES AT RISK	JC	2012	SOCIAL
LOCALS FIND STRANGE BIRD LIVING IN MANGROVE SWAMP	FIRST EDITION	2012	SOCIAL
GANG TRAFFICKED ON THE BANKS OF A MANGROVE SWAMP	CORREIO PB	2012	SOCIAL
RURALISTS PROMOTE CHANGES TO THE NEW FORESTRY CODE	JC	2012	GOVERNMENT
ENVIRONMENTAL WORKERS PUBLICISE MANIFESTO ON MANGROVES	FIRST ED.	2012	GOVERNMENT
VILLAGE IS BUILT IN MANGROVE AREA	GAZETTE	2012	ECONOMIC
MANGROVE LANDFILL SURVEYS	FIRST ED.	2012	GOVERNMENT
THE DESTRUCTION OF MANGROVES	DIÁRIO PE	2013	SOCIAL
ACTION TO REMOVE RUBBISH FROM MANGROVE AREA IN IMBIRIBEIRA	JC	2013	GOVERNMENT
AFTER RAIN, CRABS FLEE THE MANGROVES.	24H	2013	SOCIAL
LEGISLATIVE ASSEMBLY AUTHORISES MANGROVE CLEARING	DIÁRIO PE	2013	GOVERNMENT
MANGROVE LANDFILL	DIÁRIO PE	2013	ECONOMIC
CRAB HUNTING BANNED	DIÁRIO PE	2013	GOVERNMENT
CRABS FLEE THE MANGROVES AND DIE ON THE BEACH	24H	2013	GOVERNMENT
CRABS SEIZED AT A FAIR AND RETURNED TO THE MANGROVES	JC	2013	GOVERNMENT
BAN ON CATCHING UÇA CRAB BEGINS IN PE	DIÁRIO PE	2013	GOVERNMENT
ENVIRONMENTAL COMPENSATION	DIÁRIO PE	2013	GOVERNMENT
VIA MANGUE CONCRETE INVADES MANGROVE SWAMP	DIÁRIO PE	2013	ECONOMIC
RAISING STUDENTS' AWARENESS OF LITTER IN THE MANGROVES	LEAF	2013	SOCIAL
MISSING MAN'S BODY FOUND IN MANGROVE SWAMP	CORREIO PB	2013	SOCIAL

RUBBISH PILES UP IN THE SANTA CRUZ CANAL	DIÁRIO PE	2013	SOCIAL
SUBDIVISION OCCUPIES MANGROVE AREA	JC	2013	ECONOMIC
MANGROVES OF VILA VELHA	DIÁRIO PE	2013	SOCIAL
MANGROVES BETWEEN PORTO AND MARACAÍPE ONCE AGAIN OCCUPIED	JC	2013	ECONOMIC
MANGROVE DWELLERS LIVE IN THE TIME OF JOSUÉ DE CASTRO	DIÁRIO PE	2013	SOCIAL
CHILD DIES IN MANGROVE SWAMP	CORREIO PB	2013	SOCIAL
WALL HID MANGROVE DEFORESTATION	DIÁRIO PE	2013	SOCIAL
NATURE WASTED IN OLINDA	JC	2013	SOCIAL
NEITHER MANGROVE NOR INDIE	GAZETTE	2013	SOCIAL
RESTING OVERLOOKING THE SEA	JC	2013	ECONOMIC
OPERATION RIO - WASTE IN MANGROVES	FIRST ED.	2013	GOVERNMENT
GROUPER FISH BURIED IN SUAPE MANGROVE SWAMP	JC	2013	SOCIAL
PROJECT PLANTS MANGROVE SEEDLINGS ON THE BANKS OF THE JABOATÃO RIVER	JC	2013	SOCIAL
ECOLOGICAL TOURISM IN THE MANGROVES	FIRST ED.	2013	ECONOMIC
A CRY FOR DIGNITY AND RESISTANCE	JC	2013	SOCIAL
VIA MANGUE ADVANCES	JC	2013	ECONOMIC
THE MANGROVE SWAMP ADVANCES AND THE MANGROVE SWAMP SHRINKS	DIÁRIO PE	2013	ECONOMIC
VISIT TO THE MANGROVE FOREST	DIÁRIO PE	2013	SOCIAL
TEENAGER STRANGLED AND THROWN INTO MANGROVE SWAMP	24H	2014	SOCIAL
AFTER 365, CRABS RETURN TO THE MANGROVES OF THE GUARATUBA RIVER.	24H	2014	SOCIAL
DEFORESTED MANGROVE AREA IN JIQUIÁ BEGINS TO BE RECOVERED	DIÁRIO PE	2014	SOCIAL
MANGROVE AREA ONCE AGAIN OCCUPIED	GAZETTE	2014	SOCIAL
MANGROVE AREAS IN MARAGOGI ARE IN GREAT DANGER	GAZETTE	2014	ECONOMIC

Title	Source	Year	Category
ATTENTION TURNED TO THE ISLAND OF GOD	JC	2014	GOVERNMENT
MANGROVE-DERIVED BIOPRODUCTS	24H	2014	SOCIAL
FENCES TO PRESERVE THE JIQUIÁ MANGROVE SWAMP	JC	2014	GOVERNMENT
CONSTRUCTION IN MANGROVE AREAS	24H	2014	ECONOMIC
IRREGULAR CONSTRUCTIONS ADVANCE IN MANGROVE AREA	GAZETTE	2014	ECONOMIC
BOY'S BODY FOUND IN MANGROVE SWAMP	GAZETTE	2014	SOCIAL
BODY FOUND IN MANGROVE SWAMP IN BARRA DE SÃO MIGUEL	24H	2014	SOCIAL
FACING THE SEA, WITH YOUR BACK TO THE LAW.	JC	2014	GOVERNMENT
WORLD MANGROVE DAY	DIÁRIO PE	2014	SOCIAL
DAY TO TAKE CARE OF THE RIVER AND THE MANGROVES	JC	2014	GOVERNMENT
DEVELOPMENTS IN MANGROVE AREAS	24H	2014	ECONOMIC
MANGROVE SPECIES	DIÁRIO PE	2014	SOCIAL
MANGROVE GAMBOAS	GLOBO	2014	ECONOMIC
MAN DROWNS WHILE FISHING IN MANGROVE SWAMP	GLOBO	2014	SOCIAL
INVASIONS NEAR MANGROVES	JC	2014	ECONOMIC
ALLIGATOR WAS RELEASED IN A MANGROVE SWAMP	JC	2014	SOCIAL
RUBBISH IN THE MANGROVES	CORREIOPB	2014	GOVERNMENT
LARGEST URBAN MANGROVE SWAMP	JC	2014	SOCIAL
JIQUIÁ MANGROVE STILL UNDER THREAT	JC	2014	SOCIAL
IGARASSU MANGROVES	DIÁRIO PE	2014	SOCIAL
MANGROVES ARE BEING SUPPRESSED AND LANDFILLED	24H	2014	GOVERNMENT
MANGROVES THREATENED IN IGARASSU	JC	2014	SOCIAL
JIQUIÁ MANGROVE SWAMP, OCCUPIED BY IRREGULAR HOUSING.	DIÁRIO PE	2014	ECONOMIC
PUBLIC PROSECUTOR'S OFFICE INVESTIGATES ENVIRONMENTAL CRIMES	GAZETTE	2014	GOVERNMENT
RESIDENTS APPREHENSIVE ABOUT THE PRESENCE OF MANGROVE CAIMANS	LEAF	2014	SOCIAL

Title	Source	Year	Category
MPPE RECOMMENDS MEASURES AGAINST DEFORESTATION IN JIQUIÁ MANGROVE SWAMP	JC	2014	GOVERNMENT
MUNICIPALITY HAS LOST 20% OF ITS MANGROVES IN RECENT YEARS	GAZETTE	2014	ECONOMIC
IN PB, MORE THAN 1,500 UÇA CRABS ARE RESCUED AND TAKEN TO THE MANGROVES.	PB JOURNAL	2014	SOCIAL
MANATEE WATCHING DRIVES COMMUNITY TOURISM	GAZETTE	2014	SOCIAL
ORIGINS OF THE MANGUEBIT MOVEMENT	LEAF	2014	SOCIAL
MANGROVE OYSTERS	GLOBO	2014	ECONOMIC
TO SAVE THE JIQUIÁ MANGROVE SWAMP	JC	2014	SOCIAL
MANATEE IS RELEASED AND RETURNS TO THE MANGROVES	24H	2014	SOCIAL
FISHERMAN FINDS BODY IN MANGROVE SWAMP	CORREIO PB	2014	SOCIAL
TYRES IN THE MANGROVE	DIÁRIO PE	2014	SOCIAL
POPULATION BUILDS IN MANGROVE AREA	GAZETTE	2014	SOCIAL
CITY HALL ACCEPTS RECOMMENDATION TO PRESERVE MANGROVE AREA	LEAF	2014	GOVERNMENT
CITY HALL BEGINS FENCING OFF JIQUIÁ MANGROVE SWAMP	LEAF	2014	GOVERNMENT
RECIFE TO ADOPT MEASURES AGAINST MANGROVE DEFORESTATION	DIÁRIO PE	2014	GOVERNMENT
MANGROVE REPLANTING ON THE AGENDA OF ALAGOAS GOVERNMENT MEETING	24H	2014	GOVERNMENT
SOS MANGROVE AND ENVIRONMENT	JC	2014	SOCIAL
LIFE ON THE CAPIBARIBE RIVER	DIARIO DE PE	2015	SOCIAL
BUILDINGS IN MANGROVE AREAS SHOULD BE DEMOLISHED	24H	2015	SOCIAL
ESPAÇO CIENCIA APPEALS TO PRESERVE MANGROVES	JC	2015	SOCIAL
INSPECTION CHECKS IRREGULAR OCCUPATIONS IN BARRA NOVA	24H	2015	SOCIAL
CESSPITS BUILT WITHOUT GUIDANCE CONTAMINATE WATER	GAZETTE	2015	SOCIAL

Title	Source	Year	Category
CLEANING UP THE MANGROVES ON THE BANKS OF THE CAPIBARIBE RIVER	DIARIO DE PE	2015	SOCIAL
MORE THAN 800 TURKEY CRABS WERE SEIZED OVER THE WEEKEND	PB JOURNAL	2015	ECONOMIC
CAPIBARIBE MANGROVE IS OVERTURNED IN THE TOWER	JC	2015	ECONOMIC
mangroves are more VULNERABLE	GAZETTE	2015	SOCIAL
JOINT EFFORT TO COLLECT MANGROVE RUBBISH IN OLINDA	JC	2015	COCIAL
ON WORLD MANGROVE PROTECTION DAY, SECRETARIAT PROMOTES CULTURAL AND EDUCATIONAL ACTIVITIES	DIARIO DE PE	2015	SOCIAL
MANGROVE NATURAL PARK CRIES OUT FOR HELP AMID RUBBISH	DIARIO DE PE	2015	SOCIAL
PERNAMBUCO LOSES 4.8 THOUSAND HECTARES OF CONSERVATION AREAS IN NINE YEARS	JC	2015	POLITICAL
FISHERMEN FIGHT TO CONSERVE MANGROVES IN RECIFE	JC	2015	POLITICAL
ENVIRONMENTAL PRESERVATION GENERATES PROFIT FOR CRAFTSWOMEN IN BARRA DE MAMANGUAPE	PB JOURNAL	2015	ECONOMIC
PROGRAMME TO REMEMBER WORLD MANGROVE DAY	JC	2015	SOCIAL
CAPIBARIBE RIVER'S HEALTH IS SHAKEN BUT IT'S ALIVE	DIARIO DE PE	2015	SOCIAL
CAPIBARIBE RIVER: PRESERVATION IS KEY	DIARIO DE PE	2015	SOCIAL
UNCONTROLLED: DISORDERLY GROWTH ON THE NORTH COAST DESTROYS THE ENVIRONMENT	GAZETTE	2015	ECONOMIC
MEDICINAL USE OF RED PROPOLIS FROM ALOGOAS ADVANCES AND ATTRACTS SCIENTISTS FROM AROUND THE WORLD	GAZETTE	2015	SOCIAL
RECIFE CITY HALL ORGANISES JOINT EFFORT TO CLEAN UP THE BANKS OF THE CAPIBARIBE RIVER	DIARIO DE PE	2017	SOCIAL
THE RELATIONSHIP BETWEEN THE PEOPLE OF RECIFE AND THE MANGROVES OF RECIFE	DIARIO DE PE	2017	SOCIAL
CLEAN-UP ACTION ON THE BANKS OF THE CAPIBARIBE RIVER	DIARIO DE PE	2017	SOCIAL

Title	Source	Year	Category
	GAZETTE	2017	SOCIAL
AGGRESSION CAUSES STATE TO LOSE 14% OF MANGROVES IN 17 YEARS			
THE ENVIRONMENT IS A NURSERY AND HOME TO MANY SPECIES	GAZETTE	2017	SOCIAL
LEARNING TO LIVE WITH THE ECOSYSTEM	DIARIO DE PE	2017	SOCIAL
SCHOOL BOAT PROMOTES REFLECTION ON SELECTIVE COLLECTION ON BOARD THE CAPIBARIBE RIVER	DIARIO DE PE	2017	SOCIAL
CRAB HUNTING BANNED DURING CLOSED SEASON	PB JOURNAL	2017	ECONOMIC
CAPIBARIBE: A RIVER WAITING FOR ITS CITY	DIARIO DE PE	2017	ECONOMIC
MANGROVE DAY HAS CHILDREN AND YOUNG PEOPLE COLLECTING RUBBISH IN PINA	JC	2017	SOCIAL
WORLD MANGROVE DAY REMEMBERED WITH EVENT AT RIOMAR	DIARIO DE PE	2017	SOCIAL
MEETING THIS FRIDAY DISCUSSES CONSERVATION OF THE NORTH COAST OF PE	LEAF	2017	SOCIAL
MANGROVES LOSE 20 PER CENT IN 15 YEARS, SAYS SURVEY	JC	2017	SOCIAL
MANGROVES AND THE COAST ARE A NURSERY IN THE MIDST OF THE URBAN CHAOS OF THE RMR	LEAF	2017	SOCIAL
RESIDENTS WORK ON PROJECTS TO MAINTAIN THE NATURAL BEAUTY IN AL AND PE	LEAF	2017	SOCIAL
A NEW NEIGHBOURHOOD IS BORN - INVADED	DIARIO DE PE	2017	ECONOMIC
ON MANGROVE DAY, MUTIRAO CLEANS RIVERBANK IN RECIFE	GLOBO	2017	SOCIAL
THE FUTURE OF THE CITY PULSES IN THE MANGROVES	DIARIO DE PE	2017	SOCIAL
NGO GATHERS FISHERMEN TO COLLECT RUBBISH IN THE CAPIBARIBE RIVER	TV JOURNAL	2017	SOCIAL
NGOS LAUNCH PROJECT TO SAVE ITAPISSUMA'S MANGROVES	JC	2017	SOCIAL
MANGROVE PARK REQUIRES CARE	DIARIO DE PE	2017	GOVERNMENT
CAPIBARIBE RIVER RECEIVES CLEAN-UP AND ENVIRONMENTAL AWARENESS CAMPAIGN	DIARIO DE PE	2017	SOCIAL

REFERENCES

ALMEIRA, F. C. **Manguezais Aracajuanos: coexisting with devastation**. Editora Massangana, Recife. 2010.

ALMEIDA JR., A. R.; ANDRADE, T. N. Publicidade e Ambiente: alguns contornos. In: **Ambiente & Sociedade**. Campinas, v. X, n.1. 2007. p. 107-120.

ALVES, Jane Magali Rocha. The role of the media in environmental information. In: **XXV Brazilian Congress of Communication Sciences**. Salvador/BA. 2002.

ANDRADE, T. N. Environmental conservation and the media: new trends. In: **LUCHIARI**, Maria (eds). Heritage, nature and culture. São Paulo: Cornacchia. 2007. p. 163-170.

ARAÚJO, D. S. D.; MACIEL, N. C. The mangroves of Guanabara Bay. In: **Cadernos FEEMA** - Technical Series. Rio de Janeiro. 1979. n.10/79, p. 1-113.

ARAÚJO, F. B.; GONDIM, C. J. E.; FIGUEIREDO, M. F. K.; COLÓSIO, A. C.; BARRIGA, R. H. M. P.; GONDIM, C. H. M.; GUEDES, P. R. Actions and results of the Soure mangrove peoples project: Life between the land and the sea (ajó). Petrobrás Environmental Programme & NGO Novoscurupiras. In: **IV Regional Meeting on Environmental Education in Mangrove Areas**. Recife. 2005. 29p.

BARROS, H. M.; LEÇA - ESKINAZI, E.; MACEDO, S. J.; LIMA, T. **Gerenciamento Participativo de Estuários e Manguezais**. Editora Universitária UFPE, Recife. 2000.

BASÍLIO, T. H.; COSTA, T. E. S.; BEZERRA, L. N.; FERREIRA, B. A.; COLARES, L. P.; NETO, A. S. R. ; TAVARES, C. E. A.; GOMES, S. O. Reforestation of an urban mangrove area as an environmental education practice in the Cocó Ecological Park, Fortaleza - CE. In: **IV Regional Meeting on Environmental Education in Mangrove Areas**. Recife. 2005. 31p.

BUENO, Wilson da Costa. **Communication, journalism and the environment: theory and research**. São Paulo, Marajoara Editorial, 2007.

CARVALHO, C. O. G.; TONI, F. Analysing mainstream media coverage of deforestation in the Amazon. In: **Media and the**

environment: studies and essays. ALMEIDA JUNIOR, Antonio Ribeiro de (org.). São Paulo: Hucitec, 2009, p.225-255.

CAVALCANTI, C. (org,) **Meio Ambiente, Desenvolvimento Sustentável e Políticas Públicas**. São Paulo, 1997.

CINTRÓN, G.; SCHAEFFER-NOVELLY, Y. **Introducción a La ecologia del manglar**. Montevideo, ROSTLAC/UNESCO, 1983. 109 p.

CPRH. **Socio-environmental diagnosis of the north coast of Pernambuco**. Recife: State Agency for the Environment and Water Resources, 2003. 214 p.

COELHO JR, C.; NOVELLI, Y. Theoretical and practical considerations on the impact of shrimp farming on Brazilian coastal ecosystems, with emphasis on the mangrove ecosystem. In: **Mangroove 2000: Sustainability of estuaries and mangroves: challenges and perspectives**. Full papers... (CD-ROM) Recife: Federal Rural University of Pernambuco, 2000. 9 p.

COLETTO, L. H.; DALA VECHIA, G. S.; DIAS, A. S.; MAGRINI, C.; KOLINSKI MACHADO, F. V.; ROSA, L. R.; SORAES, M. C. & AMARAL, M. F. The environment on the agenda: how Environmental Journalism and *Civic Journalism are* practised in the newspaper Diário de Santa Maria. In: **XI Congress of Communications Sciences in the Southern Region**. Novo Hamburgo, 2010.

CORREIA, M. D. & SOVIERZOSKI, H. H. **Marine ecosystems: reefs, beaches and mangroves**. Maceió: EDUFAL, 2005.

CUNHA, A. Lessons in the Chico Science mangrove, Espaço Ciência, Olinda - PE. In: **Mangroove 2000: Sustainability of estuaries and mangroves: challenges and perspectives**. Complete works... (CD- ROM) Recife: Federal Rural University of Pernambuco, 2000. 5 p.

DIEGUES, A. C.; **Human ecology and planning in coastal areas**. São Paulo: NUPAUN-USP, 1996.

ESTEVES, P. C. D.; POLLERY, R. C. G.; FERNANDES, S. L. V. Seasonal fluctuation of water quality in the Maricá Lagoon, Rio de Janeiro. In: **Brazilian Plankton Meeting**, 4, Recife, 1990. Proceedings... Recife: Federal University of Pernambuco, 1991. p.

485-499.

FERNANDES, A. J.; PERIA, L. C. S. Characteristics of the environment. In: SCHAEFFER-NOVELLI, Y. **Manguezal: ecossistema entre a Terra e o mar**. São Paulo: CaribbeanEcologicalResearch, 1995. p. 13-15.

FERNANDES, F. A. M. The role of the media in the defence of the environment. Available at: < **http://site.unitau.br/scripts/prppg/humanas/download/opapelmi dia- N2-2001.pdf** > Accessed on: 02 January 2015.

FERNANDES, R. T. V. **Recovery of Mangroves**. Rio de Janeiro: Interciência, 2012.

FERREIRA, D. F.; SAMPAIO, F. E. ; SILVA, R. V. C.; MATTOS, S. C. **Socio-environmental impacts caused by irregular occupations in areas of environmental interest - Goiânia/GO**. Article (Postgraduate Programme in Environmental Management). Catholic University of Goiás, 2005.

FREITAS, H. M. R.; CUNHA, M. V. M.; Jr., & MOSCAROLA, J. Aplicação de sistemas de software para auxílio na análise de conteúdo. In: **Revista de Administração da USP**, *v. 3 n. 32.1997.* *p.*97-109.

FREY, K. The role of the press in environmental policy. In: **Revista de Human Sciences - Themes of Our Century**. Federal University of Santa Catarina. Philosophy and Human Sciences Centre. v.1, n.32. Florianópolis: Editoria da UFSC, 2002. p. 293-319.

FUSER, I. (org.) **The Art of Reporting**. São Paulo: Scritta, 1996.

HISSA, C. E. V. **Saberes Ambientais: desafios para o conhecimento disciplinar**. Belo Horizonte: UFMG, 2008.

JACOBI, P. Environmental Education, Citizenship and Sustainability. In: **Cadernos de pesquisa**. n° 118, 2003. p. 189-205.

JOHN, L. Press, Environment and Citizenship. In: **Ciência & Ambiente**. Santa Maria, UFSM. 2001. p. 87-94.

MACIEL, N.C. Some aspects of mangrove ecology. In: **CPRH**, 1991. Alternatives for the use and protection of mangroves in the Northeast.
Recife, Companhia Pernambucana de Controle da Poluição

Ambiental e de Administração dos Recursos Hídricos. Technical Publications Series, N. 003, 9-37.

MAIA, L.P.; LACERDA, L.D.; MONTEIRO, L.H.U.; SOUZA, G.M. **Atlas dos manguezais do Nordeste do Brasil: Avaliação das áreas de manguezais dos Estados do Piauí, Ceará, Rio Grande do Norte, Paraíba e Pernambuco**. Fortaleza: SEMACE, 2005.

MASSUCHIN, M. G.; CERVI, E.U. Environmental policies in the Gazeta do Povo newspaper: how government environmental actions are covered. 2010. Available at: **<http://www.periodicos.ufsc.br/index.php/jornalismo/article/vie w/15 129>**. Accessed on: 01 October 2014.

MENEZES, F. P. D. Minas Gerais press and environmental issues: genres, agendas and framings. In: **Revista de Artes e Humanidades**. n.7. 2011. 2p.

MOZATTO, A.R.; GRZYBOVSKI, D. Content Analysis as a Technique for Analysing Qualitative Data in the Field of Administration: Potential and Challenges. In: **RAC**, Curitiba, v.5, 731-747 p. 2011

NORDI, N. A produção dos catadores de caranguejo-uçá (*Ucides cordatus*) na região de Váreza Nova, Paraíba-Brasil. In: **Revista Nordestina de Biologia**. n.9. 1994. p. 71-77.

OLIVEIRA, F. Democracy, environment and journalism in Brazil. In: **Communication and the Environment**. São Paulo: Edusp, 1996.

OLIVEIRA, J. A. **Percepção ambiental sobre o manguezal por alunos e professores de uma unidade escolar pública no bairro de Bebedouro, Maceió - Alagoas**. Maceió: UFAL, 2004. 36f. Monograph (Specialisation in Biology of Coastal Ecosystems, Federal University of Alagoas).

OLIVEIRA, L. A. K; FREIRAS, R.R.; BARROSO, G.F. Mangroves: Tourism and Sustainability. In: **Caderno Virtual de Turismo**, Rio de Janeiro. Volume 5, n.3. 2005

OLMOS, F. & SILVA E SILVA, R. **Guará, Ambiente, Flora & Fauna dos Manguezais de Santos-Cubatão**. Publisher: Empresa das Artes, SP. 2003.

PARK, R. News as a form of knowledge: a chapter in the sociology of knowledge. IN: **ESTEVES, João Pissarra (org.) Comunicação e**

Sociedade: os efeitos sociais dos meios de comunicação de massa. Lisbon: Livros Horizonte, 2002, p. 35-48.

POR, F. D. **Guia ilustrado do manguezal brasileiro**. São Paulo: USP Biosciences Institute: 1989. 34p.

RAMOS, L. F. A. **Meio Ambiente e Meios de Comunicação**. São Paulo: Annablume, 1995.

REIGOTA, M. O Estado da Arte da Pesquisa em Educação Ambiental no Brasil. In: **Pesquisa em Educação Ambiental**, vol.2, n.1. p. 33-66. 2007.

ROCHA, E. A.; PEREIRA, M. G.; PEPE, A. Environmental education in mangrove areas of Canavieiras, Bahia, Brazil: Views and knowledge of everyday life. In: **2nd EREBIO - Regional Biology Teaching Meeting**, João Pessoa, 2006. Proceedings... João Pessoa, UFPB. 2006. p. 156-158.

ROLNIK, R.; KLINK, J. Economic growth and urban development: why are our cities still so precarious? In: **Novos Estudos - CEBRAP**. n. 89. São Paulo. 2011.

SILVA, A.R.C. The **media's approach to sustainable local development: the case of the Pernambuco Pharmaceutical Complex**. 2011.
210f. Dissertation (Professional Master's Degree in Sustainable Local Development Management) - Pernambuco School of Administration Sciences, Recife.

SCHAEFFER-NOVELLI, Y. **Profile of Brazilian coastal ecosystems, with special emphasis on the mangrove ecosystem**. São Paulo: Oceanographic Institute of the University of São Paulo, 1989.

SCHAEFFER-NOVELLI, Y. **Mangrove, ecosystem between land and sea**. São Paulo: CaribbeanEcologicalReseaech, 1995.

SCHAEFFER-NOVELLI, Y. The ecological and socio-economic role of mangroves. In: **CAMPOS, A. A. *et al.* (coords). The coastal zone of Ceará: Diagnostics for integrated management**. Fortaleza: AQUASSIS. p. 46-47, 2003.

SCHAEFFER - NOVELLI, Y. Current Situation of the Group of Ecosystems: "Mangrove, Marisma and Apicum", Including the Main Vectors of Pressure and Perspectives for its Conservation and

Sustainable Use. São Paulo.
**Available<http://www.anp.gov.br/ibamaperfuracao/refere/mang
ueza l_marisma_apicum.pdf>Accessed** on 02 January 2015.

TAVARES, C.; FREIRE, I. M. Environmental information in Brazil:
for what and for whom. In: **Perspectivas em Ciência da
Informação**. v.8, n.2.
2003.

TRIGUEIRO, A. The environment in the media age. In: **Trigueiro,
André (coord). Meio Ambiente no século 21 - especialistas
falam da questão ambiental em suas áreas de conhecimento**.
Rio de Janeiro: Sextante, 2003. p. 75-90.

VANNUCCI, M. **Os manguezais e nós: uma síntese de
percepções**. 2. ed. São Paulo: USP, 2002.

VIANA, B. C. B. Print Media: the device. In: **9th National Meeting
on Media History**. UFOP - Outro Preto, Minas Gerais. 2013. p. 1-
11.

YOKOYA, N. S. Distribution and origin. In: **SCHAEFFER-NOVELLI,
Y.
Mangroves: ecosystem between land and sea**. São Paulo:
Caribbean Ecological Research, 1995. p. 9-12.

Table of contents

Printed by Books on Demand GmbH, Norderstedt / Germany